DISCARD

Healthy HABITS™

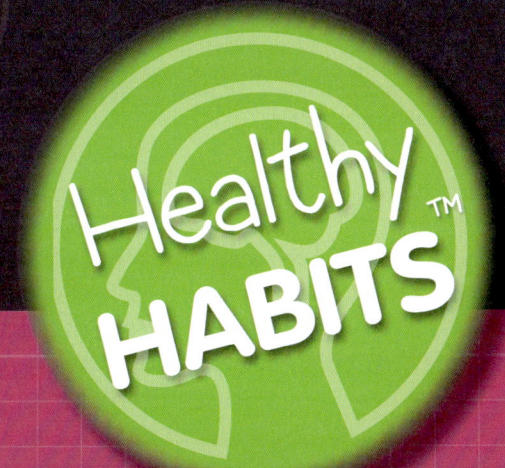

Your Beautiful Brain
Keeping Your Brain Healthy

Jeri Freedman

rosen publishing's
rosen central®

New York

Published in 2013 by The Rosen Publishing Group, Inc.
29 East 21st Street, New York, NY 10010

Copyright © 2013 by The Rosen Publishing Group, Inc.

First Edition

All rights reserved. No part of this book may be reproduced in any form without permission in writing from the publisher, except by a reviewer.

Library of Congress Cataloging-in-Publication Data

Freedman, Jeri.
 Your beautiful brain: keeping your brain healthy/Jeri Freedman.—1st ed.
     p. cm.—(Healthy habits)
 Includes bibliographical references and index.
 ISBN 978-1-4488-6952-7 (library binding)
 1. Brain—Care and hygiene. I. Title.
 QP376.F668 2013
 616.8'045—dc23

2011052009

Manufactured in the United States of America

CPSIA Compliance Information: Batch #S12YA: For further information, contact Rosen Publishing, New York, New York, at 1-800-237-9932.

CONTENTS

Introduction 4
CHAPTER 1 Your Brain 6
CHAPTER 2 Keeping Your Brain Physically Healthy 18
CHAPTER 3 Keeping Your Brain Mentally Healthy 29
CHAPTER 4 Avoiding Alcohol, Tobacco, and Drugs 36
CHAPTER 5 Exercising Your Brain 43
Glossary 49
For More Information 52
For Further Reading 56
Bibliography 58
Index 62

Introduction

The brain is a complex organ that is the control center of the nervous system. The brain controls all the functions of the body, including the functioning of organs and the movement of limbs. It is also responsible for processing emotions, thoughts, memories, and information from the senses.

Maintaining a healthy brain is critical. Keeping the brain healthy enhances your ability to concentrate, learn, and remember. Developing healthy habits will maximize your ability to think and process information and will increase your chances for success in life. On the other hand, developing unhealthy habits can harm the brain, and it can be difficult—sometimes impossible—to repair the damage.

Throughout your teens and early twenties, the brain is still growing and maturing. For this reason, forming positive habits at this time ensures that your brain develops in the best way possible. The healthy habits you start now will serve you well throughout your life. It is necessary to keep your brain healthy, not only through good physical habits but also through mental exercise. It is also important not to allow negative habits, such as poor nutrition, drugs, or lack of sleep, to undermine your brain's health.

This book explains the habits that will help you keep your brain healthy and maximize its potential. In recent years, scientists have discovered a great deal of new information about how nutrition, exercise, drinking, drugs, smoking, and other factors affect the brain. This book uses this new research to explain what you can do to keep your brain healthy. It also explores how bad habits can affect the brain. Protecting your brain—and doing what is necessary to keep it functioning well—will enhance your quality of life now and as you age.

Developing habits that keep your brain healthy will help you deal with tasks like doing homework now, as well as other mentally demanding tasks as you go through life.

Chapter 1

Your Brain

Your brain controls not only what you think and feel but also the functioning of your body. The brain is a major organ of the central nervous system, which consists of the brain, the spinal cord, and the peripheral nerves. The peripheral nerves extend from the spinal cord to every part of the body. Instructions—in the form of electrical signals—travel from the brain to the spinal cord and along the peripheral nerves. Data, or information—also in the form of electrical signals—travels from the peripheral nerves up the spinal cord to the brain.

The brain receives information from sensory nerves in the nose, tongue, ears, eyes, and skin. It processes this information to provide you with sensory data in the form of smell, taste, hearing, sight, and touch. The brain also controls the muscles. It regulates the body's internal functions, such as heart rate, breathing (respiration), blood pressure, and body temperature. It encodes and stores your memories and is responsible for your feelings and emotions.

The basic cell of the nervous system is the neuron, also known as a nerve cell. The brain contains billions of neurons, which send electrical signals to communicate with and control the muscles and other organs in the body. A second type of cell in the nervous system, known as a glial cell, serves to support and nourish the neurons.

Your Brain

Spinal Cord

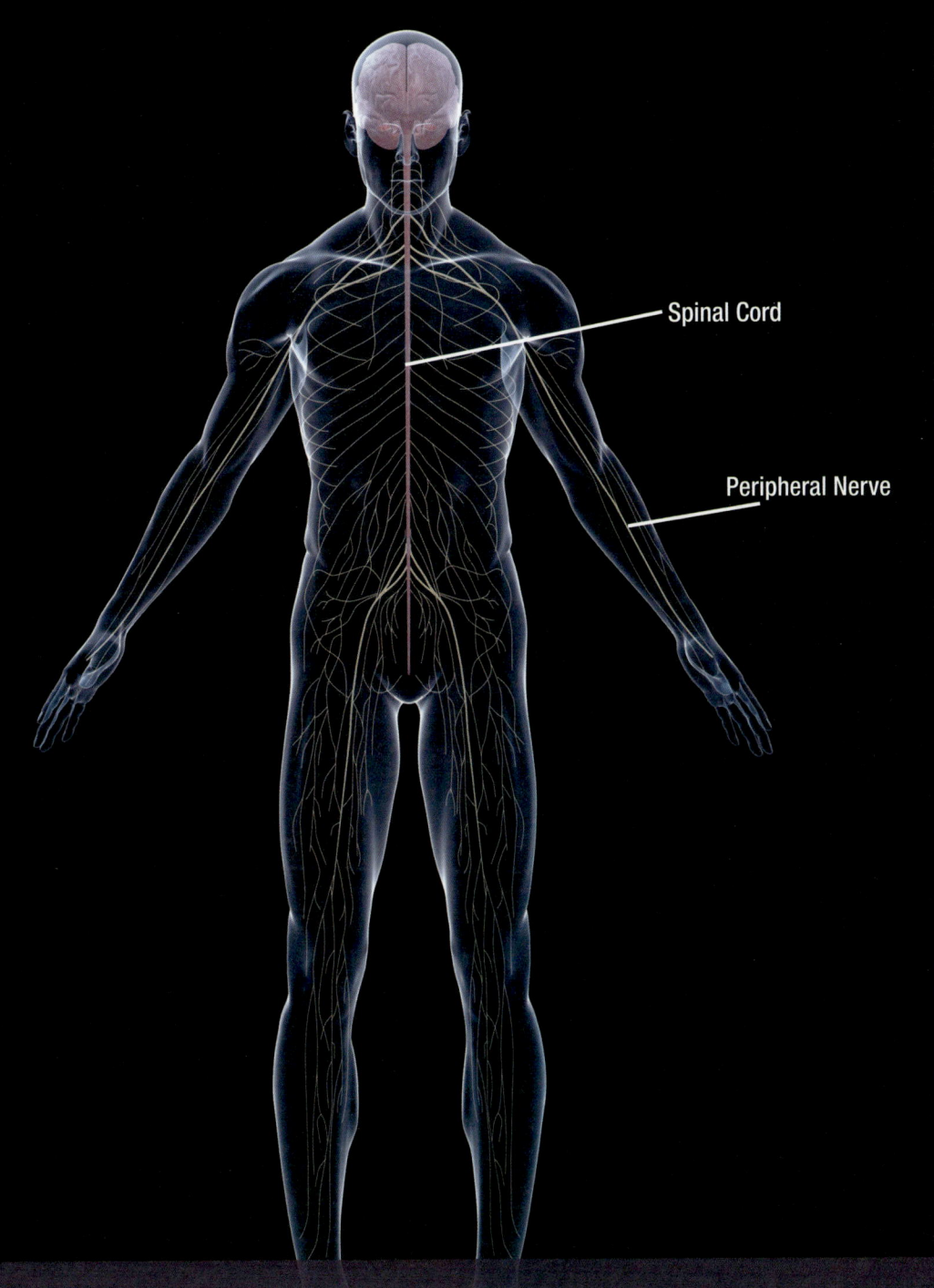

Peripheral Nerve

The nerves extend from the brain down the spinal cord and from the spinal cord to other parts of the body, such as the arms and legs.

Your Beautiful Brain: Keeping Your Brain Healthy

Structure and Functions of the Brain

Your brain is divided into various parts, each of which is responsible for particular functions. The different parts of the brain also work together. The major parts of the brain include the brain stem, the cerebellum, the cerebrum, and the limbic system.

The brain stem is located at the tail end of the brain and connects with the top of the spinal cord. It has three parts: the medulla, the pons, and the midbrain. The brain stem controls the basic functions that our bodies perform without conscious effort—the activities that the body carries out automatically. Among these are reflexes, blood

Human Brain - Side View

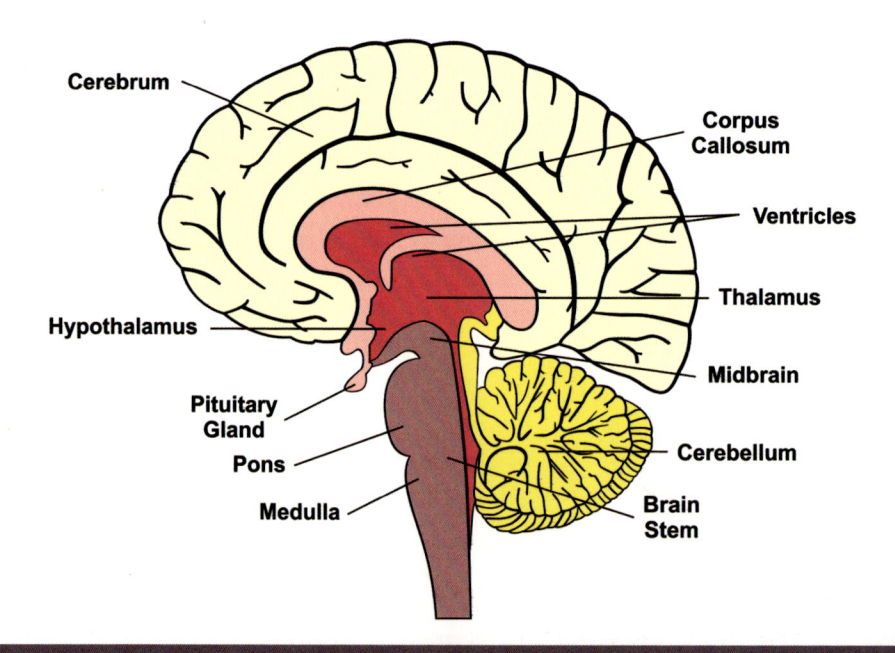

The different sections of the brain, shown in the illustration above, are responsible for specific functions but also work together so that people can perform complex tasks.

pressure, breathing, heart rate, digestion, involuntary movement, and sleep. The brain stem also controls the main sensory and motor functions of the face and neck.

The cerebellum sits at the back of the brain above the brain stem. The cerebellum receives information from the body on the positioning of limbs. It uses this information to control balance and move the limbs in a coordinated manner—when you walk, for example.

The cerebrum is the largest part of the brain, making up about 85 percent of the brain. It consists of a number of major parts that are responsible for reasoning and analyzing information. It also contains areas that process sensory data. In addition, the cerebrum controls voluntary movement.

The limbic system consists of the thalamus, hypothalamus, hippocampus, and amygdala. These structures are found deep inside the cerebrum. The thalamus regulates sleep and alertness. The hypothalamus is responsible for maintaining body temperature. It also controls hunger, thirst, and emotional responses such as anger and pleasure. The hypothalamus also controls activities by way of the pituitary gland, a pea-sized organ located at the base of the brain. This gland releases hormones that are necessary to control growth and other bodily processes. The hypothalamus sends signals to the pituitary gland, managing the release of hormones. The hippocampus controls both short-term and long-term memory. The amygdala plays a key role in emotions such as fear and anger.

The Structure of the Cerebrum

A deep groove divides the cerebrum into two halves, the right hemisphere and left hemisphere. The two hemispheres are connected by the corpus callosum, a band of fibers that runs through the space

Your Beautiful Brain: Keeping Your Brain Healthy

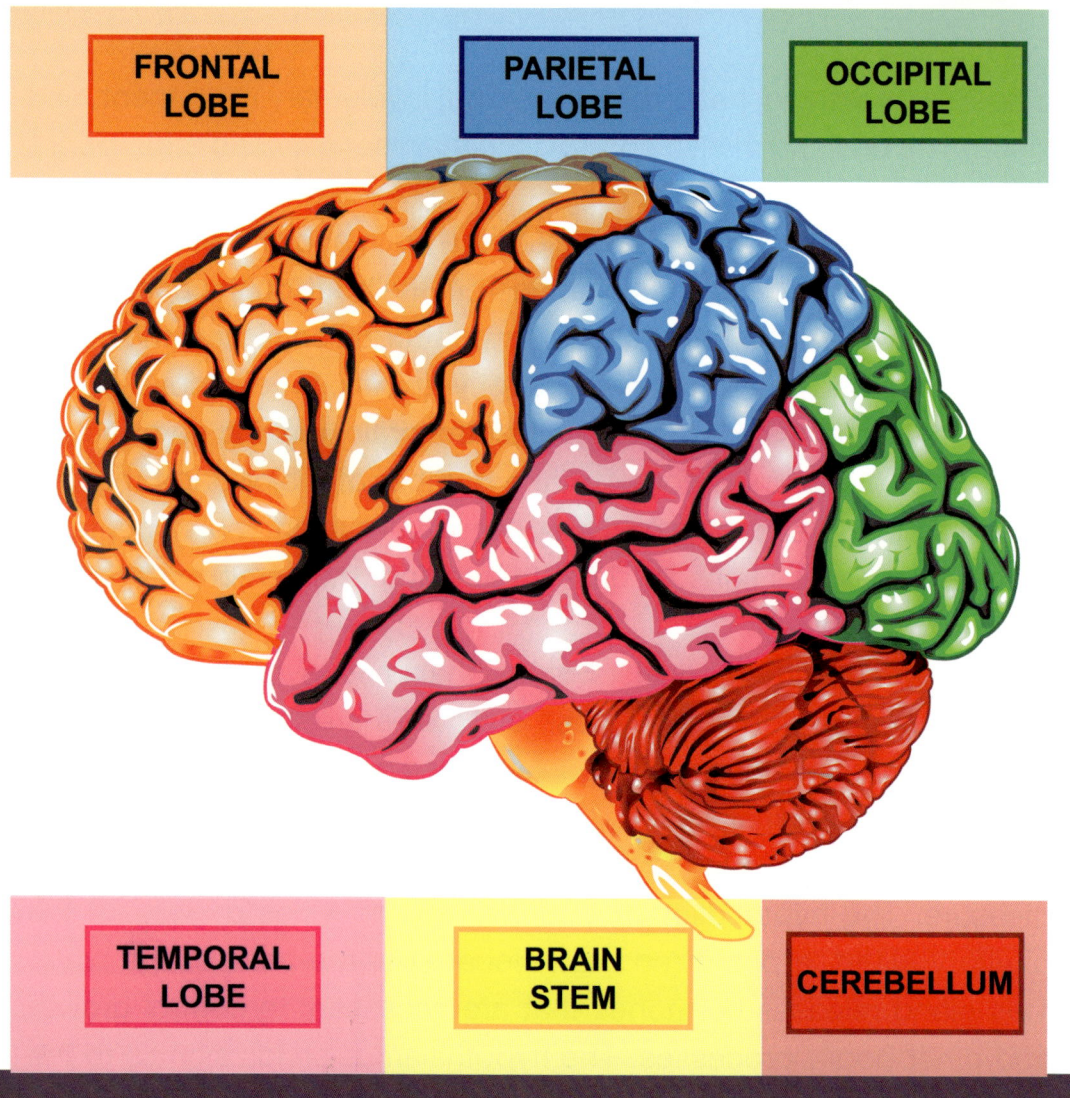

FRONTAL LOBE	PARIETAL LOBE	OCCIPITAL LOBE

TEMPORAL LOBE	BRAIN STEM	CEREBELLUM

This illustration shows the size and location of the various lobes in the brain. A second matching set of lobes exists on the other side of the brain.

between them. The layer of gray matter covering each hemisphere is known as the cerebral cortex.

The cerebrum is also divided into several major areas called lobes, including the frontal lobes, parietal lobes, temporal lobes, and occipital lobes. The lobes occur in pairs, with one of each type in the right hemisphere and one in the left. A complex set of fibers allows the right and left lobes to "talk" to each other. The lobes of the brain have the following functions:

- **Frontal lobes.** Lying directly behind the forehead, the frontal lobes are important in thought, analysis, and learning. These lobes also help control voluntary movement and motor skills. On the left frontal lobe is an area called Broca's area, which helps a person transform thoughts into words, controlling the muscles used in speech.
- **Parietal lobes.** Located behind the frontal lobes, the parietal lobes receive sensory information, such as touch, taste, and temperature, from the body. At the rear of the parietal lobe is an area called Wernicke's area. This area is critical for comprehending speech.
- **Temporal lobes.** The temporal lobes lie under the frontal and parietal lobes. These lobes receive and process auditory information such as sounds and music.
- **Occipital lobes.** Located at the back of the brain, the occipital lobes process visual information (images from the eyes).

The names of the different lobes come from the names of the main skull bones that cover them.

Structure of a Neuron

A neuron is the basic cell of the brain and nervous system. Each neuron has the following parts:

- **Cell body.** This part of the neuron contains the structures necessary to process nutrients and perform other basic functions.

Your Beautiful Brain: Keeping Your Brain Healthy

PARTS OF A NEURON

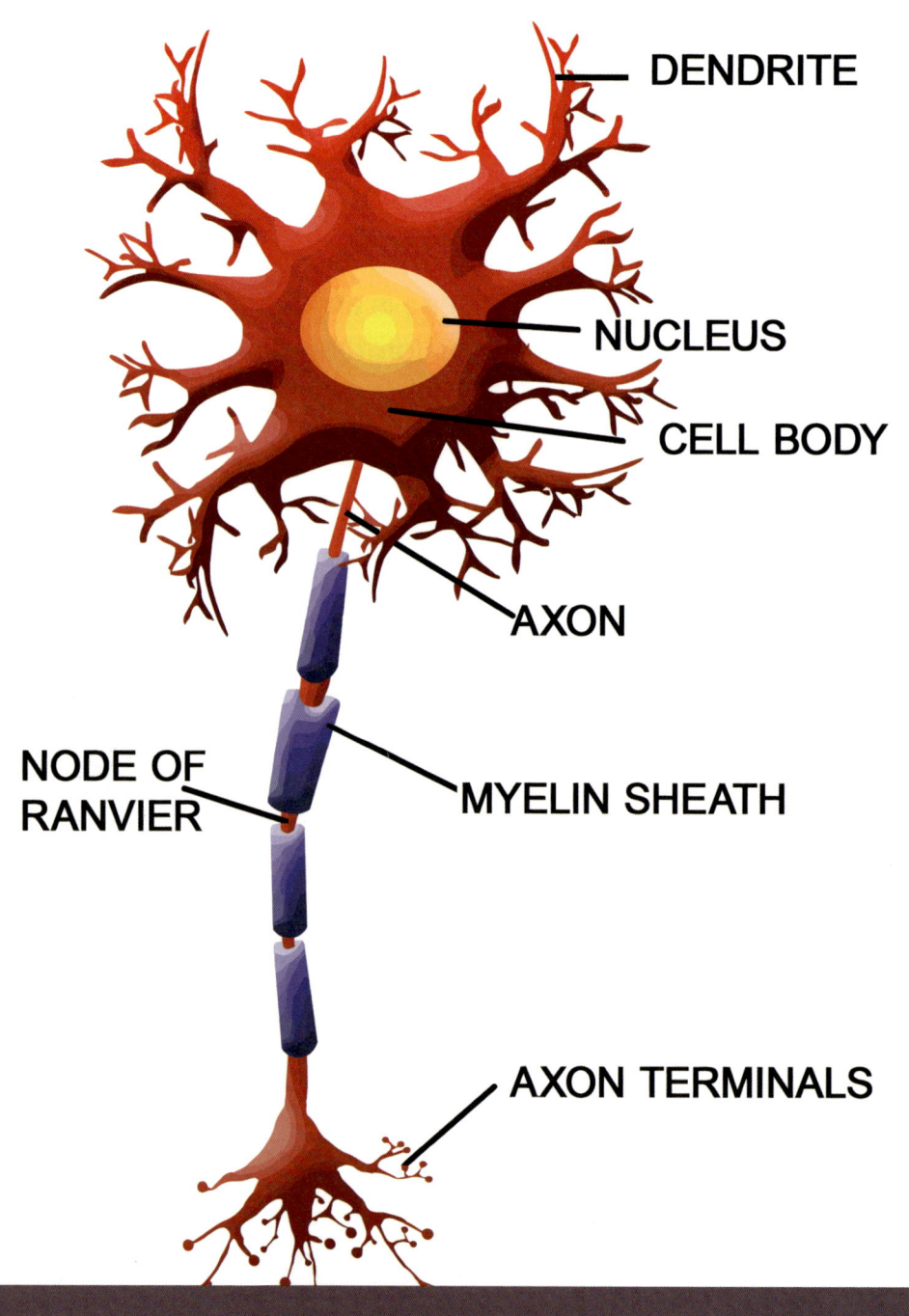

- DENDRITE
- NUCLEUS
- CELL BODY
- AXON
- NODE OF RANVIER
- MYELIN SHEATH
- AXON TERMINALS

In a neuron, electrical signals jump from gap to gap in the myelin sheath until they reach axon terminals, where they cross a space called a synapse to an adjacent neuron.

- **Axon.** The axon is a long, taillike structure extending from the cell body. It is covered by a fatty sheath, which protects it. Electrical signals travel down the axon to the synapse—the space where neurons meet.
- **Dendrites.** These hairlike fibers extend out from the cell body and receive messages from other neurons. They allow the neuron to form connections to other neurons in order to create complex networks.
- **Synapse.** This is a small gap at the end of the axon where electrical pulses are transmitted to other neurons and received from them.

Neurotransmitters are chemicals that help transmit signals from one neuron to the next. These chemicals are released from the end of the axon into the synapse. The neurotransmitters cross the synapse and attach to receptors on the receiving neuron's dendrites. The signal can then be transmitted to the next neuron. At the end of the chain, neurons can form synapses with other kinds of cells, such as muscle cells or gland cells, and can send neurotransmitters to stimulate them.

The Teenage Brain

Until the 1990s, experts believed that the brains of children, teenagers, and adults were structurally the same. They believed that the only thing that changed during children's development was that new connections formed between neurons as children learned.

Beginning in the late 1990s, this view changed radically. The technology of magnetic resonance imaging (MRI) allowed scientists to take detailed images of the brain. A series of these images could

Diseases of the Brain

Disease or injury to the brain can affect bodily functions and thinking. The brain is subject to a number of diseases, some of which are genetic. Genetic diseases result from mutations, or changes, in genes. These mutations may be either random or inherited. An example of an inherited disease is Huntington's disease, which starts in adulthood and causes brain cells to waste away and die.

Other diseases result from changes that occur with age. These are called degenerative diseases because they result from the breakdown of cells. One example is Alzheimer's disease, in which plaques (flat, raised areas of protein) and tangles of nerve fibers form in the brain and cause loss of brain function.

Another example of a brain disorder is a stroke, or a stop in blood flow to a part of the brain. Strokes can occur at any age, but the likelihood increases with age. Also known as a "brain attack," a stroke is sometimes caused by a blockage in a blood vessel that supplies blood to the brain. It can also result from the bursting of a blood vessel in the brain, causing blood to leak into the brain. Because blood and oxygen cannot reach part of the brain, a stroke can result in the death of brain cells. Strokes can cause paralysis and the loss of speech or other important functions.

Although we can't stop aging, we can affect how susceptible we are to the diseases that come with age by developing healthy habits when we are young. For example, strokes are more likely in people who have high blood pressure, a history of smoking, or arteriosclerosis (hardening of the arteries), a disease in which blood vessel walls fill with hard, fatty deposits. High blood pressure and arteriosclerosis are associated with a high-fat diet and obesity.

show the changes that take place in children's developing brains. Now it is known that when children enter puberty in their early teens, their brains undergo a growth spurt, just as their bodies do. Both imaging studies and evidence from direct examination of brains after death indicate that different parts of the brain mature at different times. It appears that the frontal lobes, the part of the brain involved in reasoning and judgment, do not completely mature until the early twenties.

In one significant study, scientists at the University of California, Los Angeles, compared MRI brain scans of twelve- to sixteen-year-olds with those of twenty-three- to thirty-year-olds. They found that the areas of the parietal and temporal lobes related to spatial, sensory, auditory, and language processing were quite mature in the teenage brain. In contrast, there was significantly less maturation in parts of the frontal lobes in the teen brains. These were the areas responsible for cognition (thinking), including perception, judgment, reasoning, and analysis. Other studies have resulted in similar findings.

In another study carried out at McLean Hospital in Belmont, Massachusetts, Deborah Yurgelun-Todd and her research team used MRI technology to study how teenagers and adults respond emotionally. They took images of the brains of adults and the brains of children ages ten to eighteen. When shown a picture of a person displaying fear, the adults correctly identified the emotion. The teenagers, however, described the expression as shocked, angry, or surprised. The brain images revealed that the adults were processing what they saw through the frontal cortex, the area of the brain that governs reason. However, the teenagers were processing emotional information through the amygdala, a deep brain structure responsible for instinctual reactions such as fear and aggression.

The amygdala activates strong, instantaneous reactions like the "fight-or-flight" response, which helps people respond quickly to danger. Thus, the teenagers responded to emotional input with a "gut" reaction. This may explain the frequently powerful and, sometimes unpredictable, emotional reactions of teenagers. In the study, the older the teenager, the more the processing of emotional information shifted to the frontal cortex area.

These discoveries go a long way toward explaining some of the behavior of teenagers. Long before imaging technology existed, people noticed the changes in maturity and behavior that take place between the teen years and adulthood. Developments in brain science are now providing an explanation.

In addition, the research results suggest the importance of healthy habits during adolescence. Because key parts of the brain are still developing, teens must begin and maintain habits that keep the brain healthy. Doing everything you can to help your brain develop well can significantly enhance your chances for success in the future.

MYTHS and FACTS

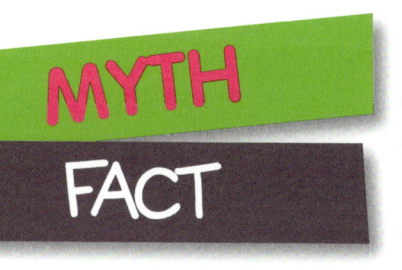

MYTH Human beings have the largest brain.
FACT Animals that are much larger than people, such as whales, have larger brains, but human beings have the heaviest brains in comparison to their body size.

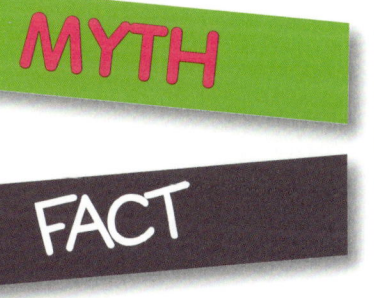

MYTH The brain doesn't grow new cells.
FACT For a long time this was thought to be true. However, recent research has shown that when the brain is damaged, one way it seeks to repair itself is by growing new cells.

MYTH Most people use only 10 percent of their brain.
FACT Imaging scans of the brain have shown that people use all of the brain, although different areas are more or less active at any given time, depending on the activities of the person.

Chapter 2

Keeping Your Brain Physically Healthy

Did you know that your lifestyle and habits can affect the structure of the cells in your brain? Your habits can also affect the neurotransmitters—the chemicals that activate neurons and control communication among them. Eating properly, exercising, maintaining a healthy weight, and avoiding alcohol all contribute to a healthy brain. You already know you must do these things to have a fit body, but you also need to do them to have a fit mind.

Nutrition for a Healthy Brain

Eating properly helps ensure that your brain has the building blocks it needs to keep the structure of brain cells intact and repair them when necessary. It also helps ensure that you do not build up unhealthy substances in your cells that could harm your brain.

The digestive system breaks down the food you eat. Substances in food are carried to the cells throughout your body by the circulatory system, which includes the heart and blood vessels. In the cells, substances from food are used to produce energy that the cells need to carry out their activities. In the body, the brain is the biggest consumer of food for energy. According to an article in *Scientific American*, the brain uses as much as 20 percent of all the energy supplied to the

Keeping Your Brain Physically Healthy

Eating a well-balanced diet ensures your brain gets the nutrition you need. Avoid fad diets and eat vegetables, whole grains, and low-fat protein to keep body and mind healthy.

body. Much of this energy comes from fat and sugar. However, eating the right types of fat and sugar is important.

Eat the Right Fats and Sugars

Eating the wrong kinds of fat can negatively affect the brain. Researchers at the University of California, Los Angeles, found that eating junk food can make brain cells more vulnerable to damage and can negatively affect intelligence and memory. The trans fats and saturated fats found in junk food and much fast food interfere with the formation of synapses, or connections between neurons. These substances also interfere with the formation of molecules needed for learning and memory. Therefore, one healthy habit you should consider adopting is eating less junk food, fast food, and other food high in trans fat and saturated fat. Products that tend to be high in trans fat include fried foods, packaged baked goods such as cookies and doughnuts, processed foods, and margarines. Foods high in saturated fat include cheeses, whole milk, ice cream, and fatty meats. You can find out whether food products contain unhealthy fats by reading the nutrition facts label on food packaging. When choosing which foods to buy, compare the labels of similar products and purchase the one with the lowest amount of saturated and trans fat.

What type of fat should you eat? Your brain needs polyunsaturated fatty acids, commonly known as omega-6 and omega-3. Increasingly, packaged foods that contain this type of fat are labeled as good sources of omega fatty acids. You can also find unsaturated or polyunsaturated fats listed on the nutrition facts labels on packaged food. The most common sources of omega-3 and omega-6 are fish, especially salmon, albacore tuna, herring, and sardines, and foods such as almonds, walnuts, and broccoli. According to Fernando Gómez-Pinilla of UCLA's Brain Research Institute and Brain Injury Research Center, a lack of omega-3 in the diet has been linked to learning problems and brain disorders. In addition, he reports that eating increased amounts of omega-3 has been demonstrated to improve memory and learning in children. Apparently, the old saying that "fish is brain food" is true.

As with fat, there are healthy and unhealthy types of sugar. Sugar can be found naturally in some foods, such as fruits, but in other foods, sugars are added as sweeteners. Added sugars are those typically found in sweets, and they have little or no nutritional value. Candy, cakes, cookies, and soft drinks often contain processed cane sugar or corn syrup. There is no difference between cane sugar and corn syrup as far as the body is concerned. Both are simple sugars. The body quickly burns simple sugars, resulting in a quick rush of energy that quickly vanishes and ultimately leads to a feeling of fatigue. Excess sugar is also converted to fat, which can lead to unhealthy weight gain. Excess weight doesn't just affect your appearance: it actually harms your brain (see the sidebar on pages 24–25). So, you should seek out foods that are low in added sugars.

The brain needs a steady source of energy, so look for slowly digested carbohydrates—foods that release sugar slowly into the bloodstream. Good carbohydrates include vegetables, fruits, and whole grains such as whole wheat. These foods contain vitamins, minerals, and fiber, and their sugars are digested and absorbed slowly. When you eat crackers, bread, rice, noodles, and other grain-based products, seek out those that say "whole wheat" or "whole grain" on the package. These products contain more fiber and nutrients than refined white bread and pasta do. In addition, snack on fruits and vegetables instead of high-fat, high-sugar foods.

Eat a Good Breakfast

Don't skip breakfast! As mentioned previously, the brain is the largest consumer of energy in the body. In order to have energy, your brain needs food. When you wake up in the morning, you haven't eaten for hours, and your body and brain are short of fuel. The brain needs glucose, or blood sugar, to produce energy—and it needs a lot of it. Research has shown that when the brain does not have an adequate supply of glucose, people have trouble learning, understanding, and remembering.

What you choose to eat for breakfast also makes a difference. Eating foods high in added sugar simply provides a short burst of energy, followed by a crash as the brain quickly runs out of fuel. So

Your Beautiful Brain: Keeping Your Brain Healthy

Eat a good breakfast with whole grains to ensure that your brain has enough energy for your morning activities. Whole grains are digested slowly, providing a steady source of energy.

avoid sugar-laden cereals, doughnuts, and toaster pastries for breakfast. Instead, eat whole grains such as oatmeal or whole-grain toast, which are broken down more slowly and provide a steady amount of energy for a long time. Whole grain bread can be used to make French toast, and whole grain waffles are also acceptable. Good sources of protein, such as eggs, low-fat cheese, peanut butter, and skim milk are good to eat as well. Fresh fruit is another healthy choice.

Eat Foods That Protect Your Brain

Get into the habit of eating foods that support and protect your brain. For example, it is beneficial to eat yogurt, eggs, and soybeans. These foods help the brain make choline, a type of neurotransmitter. The spice cumin, often used in curries, is another good source of choline. A lack of choline is associated with Alzheimer's disease, which is a form of dementia (a severe loss of mental abilities).

Another chemical necessary for a healthy brain is folic acid, also known as vitamin B_9. Folic acid helps protect neurons from damage as people age. According to a study at Wageningen University in the Netherlands, taking a large amount of folic acid improves memory in elderly people. Folic acid is found in whole grains and dark green, leafy vegetables such as spinach.

Removing the toxic by-products of energy production is also important for protecting the neurons. The blood delivers oxygen and nutrients to the body's cells. Cells contain tiny organs called mitochondria, which act like little furnaces. The mitochondria use oxygen to "burn" the nutrients to release energy. In this process, unstable oxygen molecules called free radicals are released. The free radicals react with important molecules for constructing cells, damaging them. The more calories you eat, the more nutrients the cells burn, and the more free radicals are released.

Obesity and the Brain

In order for oxygen and nutrients to reach brain cells, the heart must pump them through the body's blood vessels. For this reason, behaviors that enhance the health of the heart and blood vessels are also good for the brain. Negative health behaviors can lead to the accumulation of fat and eventually the clogging and hardening of the arteries. This can weaken the walls of the blood vessels and increase a person's risk of having a stroke (a blocked or burst blood vessel in the brain), which can cause brain damage. High blood pressure also increases the chances of a stroke.

Obesity is the condition of being extremely overweight. According to the Mayo Clinic, to be considered obese a person must be 30 percent or more above one's ideal body mass index (BMI). The BMI measures what percentage of a person's body weight is fat, and the ideal numbers are based on an individual's weight and height. According to the Centers for Disease Control and Prevention (CDC), in 2010, 33.8 percent of Americans were obese, including 17 percent of children and teens ages two to nineteen.

Obesity is closely associated with high cholesterol and high blood pressure. Cholesterol is a fatty substance that the body uses to make certain hormones. There are two major

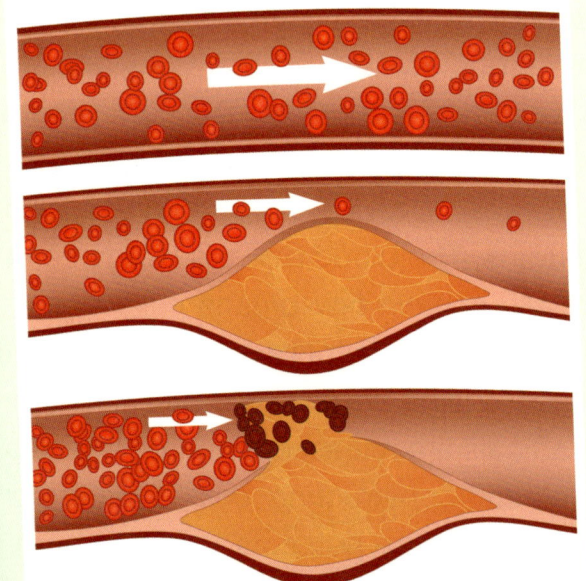

The top blood vessel is healthy. The middle artery contains fatty plaque. In the bottom artery, the plaque blocks the blood flow, leading to clotting, pressure, and possible bursting.

types of cholesterol: low-density lipoprotein (LDL) cholesterol and high-density lipoprotein (HDL) cholesterol. In addition to cholesterol, there is another form of fat that circulates in the blood—triglycerides. When LDL cholesterol and triglycerides build up in blood vessels, they reduce the space available for blood to pass through. This narrowing of the space can lead to high blood pressure. Blood pressure is a measure of the force necessary for the heart to pump oxygen-rich blood from the lungs to the rest of the body. High blood pressure can lead to a heart attack or stroke, which can have serious, and even fatal, effects.

The older you get, the harder it is to lose weight. Therefore, now is the time to develop healthy eating and exercise habits to get your weight under control. Beyond your appearance, do it for your brain!

To avoid damage from free radicals, you need to eat foods that contain antioxidants. ("Antioxidant" means "against oxygen.") The antioxidants attach to the free radicals, making them inert and harmless. Many fruits and vegetables contain antioxidants. Among the best, and tastiest, antioxidant-rich foods are blueberries, Concord grape juice (look for 100 percent juice), pomegranates, and tomatoes. Green tea is also high in antioxidants. Carrots, onions, broccoli, and spinach are also good sources. Eating more foods that maximize brain health, while eating fewer foods that are unhealthy for your brain, can actually make it easier for you to concentrate, learn, and remember. Beyond that, a healthy diet can help protect your brain and keep it functioning well as you grow older.

Exercise for Brain Health

When the brain commands a muscle to move, neurons communicate with muscle fibers across synapses. In order to make the muscle move, the nerves release a neurotransmitter called acetylcholine,

Your Beautiful Brain: Keeping Your Brain Healthy

Exercise, especially aerobic exercise, is important for brain health. Exercising regularly is easier if you get into a set routine. Asking a friend to join you can help, too.

which binds, or attaches, to receptors on the muscle fibers' surface. The muscle contracts as a result.

When muscles aren't used, the receptor structures at the synapses are disassembled. If you start exercising after a period of inactivity, your body starts to rebuild this structure. Thus, when it comes to muscle strength, it's a matter of "use it or lose it."

The heart is a muscle, and the heart is responsible for pumping blood to the brain. The brain, in turn, depends on this blood supply for oxygen and nutrients. A reduction in blood flow to the brain can result in reduced cognitive activity (mental activity, such as thinking, reasoning, and remembering). Therefore, a strong, efficient heart is needed for a well-functioning brain.

What kind of exercise should you do? The most beneficial type for the heart and brain is aerobic exercise. This kind of exercise causes a person to breathe harder and faster, raising the amount of oxygen the body takes in. Also, during aerobic exercise the heart works harder and blood flow to the brain increases. This provides the brain with more oxygen and removes more waste. Running is a good example of aerobic exercise. There are many team and individual sports that involve running, ranging from tennis to soccer. Walking is another form of exercise that is healthy for your brain. When you need to go somewhere, consider biking or walking instead of being dropped off by car (as long as it's safe to do so). The exercise will help your brain function better.

Protect Your Brain

Traumatic brain injury (TBI) occurs when the brain hits the inside of the skull because of a hard blow or when the skull is fractured (broken) and a piece of bone damages the brain. TBI can be devastating, affecting your ability to function physically and mentally for the rest of your life.

Your Beautiful Brain: Keeping Your Brain Healthy

TBI can affect a person's thinking, senses, emotions, movement, and ability to speak or understand language. It can also be fatal. In some cases, the effects of TBI can be improved with treatment. In other cases, the damage is permanent.

According to the National Institutes of Health (NIH), millions of people in the United States experience TBIs every year, and more than half of the injuries are bad enough that they require hospital treatment. Among the most common causes for TBI are sports injuries and bicycle or motorcycle accidents.

There are things you can do to prevent traumatic brain injury. Competitors should always wear protective headgear when engaging in sports such as boxing or football. If you ride a bicycle or an open vehicle, such as a motorcycle, snowmobile, or all-terrain vehicle (ATV), always wear a helmet. Accidents on bikes and open vehicles are very common. Even if you are a skilled rider, irregularities in the road or terrain and the behavior of other drivers can result in accidents. While the accidents may not be your fault, you can be injured all the same.

If you experience unconsciousness after an accident involving the head, or have blurred vision, headache, nausea, or confusion, you need to see a physician for evaluation. You may have a concussion, an injury caused by jarring of the brain, or even bleeding in the brain, which can be life-threatening. Never ignore symptoms after receiving a blow to the head. And always do all you can to protect your head from injury.

Chapter 3

Keeping Your Brain Mentally Healthy

Stress and lack of sleep can affect your ability to focus, remember, and think clearly. They can also affect your emotions and moods. This chapter examines some of the psychological and sleep-related factors that affect brain function.

Stress and Your Brain

School pressures, family issues, and problems in relationships with friends can sometimes cause stress, or strain on your mind and body. When you experience stress, the brain reacts as if you are under attack. It reacts by gearing the body up to flee or fight. To do this, it signals the adrenal glands to release a hormone called cortisol. Cortisol increases the heart rate, sending more blood to the muscles in case you need to run away or defend yourself. The problem is that exposure to cortisol can damage brain cells. When you face an emergency, it is usually short-lived. Immediately after the threat is over, the body secretes (produces) other chemicals that deactivate the remaining cortisol.

However, when you are continually stressed out about something for a long time, your body keeps releasing cortisol. The cortisol can damage cells in the hippocampus, the part of the brain responsible

Your Beautiful Brain: Keeping Your Brain Healthy

Learning ways to cope with stress will help you stay healthy. Exercise helps rebalance the "fight-or-flight" chemicals released in response to stress.

for memory and learning, and it can negatively affect your concentration, memory, and learning ability. Stress also makes it easier for toxic chemicals to get into your brain. Normally, your brain is protected by something called the blood-brain barrier, which keeps harmful substances and bacteria out of the brain. However, stress reduces the effectiveness of the blood-brain barrier.

Because of the negative effects of long-term stress, you need to develop ways of coping with stress. If you are worried or stressed out about a problem, talk to someone about it—a parent, teacher, coach, counselor, or friend. Two heads really are better than one. Others may have suggestions for dealing with the problem—solutions you may be too stressed to think of. They may also help you put the problem in perspective. Many issues are not as important in the long run as they appear in the moment. Indeed, your interpretation of the situation may not even be correct. For example, many students believe that getting low grades or tough comments from a teacher indicates that they are stupid or terrible at the subject. In fact, the teacher may actually think they have talent for the subject. He or she may simply be challenging them to develop that talent.

Other ways of coping with stress include exercise and meditation. You should not use drugs or alcohol to cope with stress. They may distract you from your problems temporarily. However, the physical stress drugs and alcohol put on your body and the fact that they impair your judgment can make your situation even worse.

Not all stress is psychological. Continuous exposure to loud noise has been shown to cause a reduction in activity in the prefrontal cortex, the part of the brain that regulates behavior. The brain interprets loud noise as a signal of danger and prepares us to flee or fight. Therefore, avoiding prolonged exposure to loud music and noise may help protect not only your ears and hearing but also your brain.

Sleep and Your Brain

People need seven to ten hours of sleep per night. Most teenagers need at least nine hours of sleep. Failing to get adequate sleep has negative effects on the brain. Not getting enough sleep can affect your concentration, hand-eye coordination, reflexes, and level of alertness. The more days you go without adequate sleep, the more hours of sleep you need to make up. This is called "sleep debt." Extreme sleep deprivation can lead to mood swings and even frightening hallucinations.

Ways to Ensure a Good Night's Sleep

Getting a good night's sleep is very important for brain function. Here are some tips for getting a good night's sleep:

- Try to go to sleep around the same time each night. This will encourage your body to prepare for sleep as that time approaches.
- Make sure the temperature in your room is comfortable. If the room is too hot or too cold, your discomfort will keep you from sleeping.
- Relax before going to bed. Read, take a hot bath or shower, listen to relaxing (not stimulating) music, meditate, or engage in other relaxing activities.
- Don't use the computer or play video games right before bed. The light from the screen can throw off your body's clock, and doing an exciting task can be stimulating, causing you to lose sleep.
- Exercise each day. Exercise helps use up stress chemicals, such as adrenaline, and releases other chemicals called endorphins that make you feel happy and relaxed. Don't exercise right before going to bed, however, as this may make you feel more, not less, awake.
- Avoid drinking coffee, energy drinks, chocolate, tea, or soda with caffeine, especially in the evening. The caffeine is a stimulant (a chemical that makes you feel more alert) and may keep you awake.
- Don't smoke. Nicotine is a stimulant.
- Avoid alcohol. Although alcohol may make you feel sleepy initially, it interferes with deep sleep, and you may wake up when it wears off in the middle of the night. Also, alcohol causes dehydration (an abnormal reduction in the amount of water in the body), which can cause headache and nausea during the night or the next day.

- Some allergy or cold pills, such as those containing pseudoephedrine, should not be taken right before bed. Pseudoephedrine is a stimulant and can keep you awake.
- If you can't fall asleep, don't keep trying to force yourself; the stress will make you feel more awake. Read, listen to music, do gentle stretches, or do something similar until you feel sleepy.

Sleep affects the physical processes in the brain. There is evidence that sleep is necessary to allow the neurons time for repair. During sleep, growth hormone is released in children and teenagers, and there is increased production of proteins, which are the building blocks of body tissues. Therefore, it is possible that sleep deprivation can affect a person's growth and development.

There are various stages of sleep, ranging from shallow to deep. One stage of sleep is called REM (rapid eye movement) sleep, because the eyes can be seen to move during this phase. People dream during REM sleep. Evidence suggests that REM sleep and deep, slow-wave, dreamless sleep are involved in the formation of memory and the retention of information learned during the day. Researchers at Harvard Medical School in Cambridge, Massachusetts, and Trent University in Ontario, Canada, examined how sleep affects learning. They trained students in some new skills, gave them a series of tests on the same day, and then had them sleep for various periods of time. Researchers examined the students again a few days later to see what they remembered. They found that during sleep, the brain practiced and stored what was learned during the day. The researchers found that the students

Your Beautiful Brain: Keeping Your Brain Healthy

The brain continues to learn when you are asleep. Getting a good night's sleep is an easy way to improve your memory and enhance your grasp of new skills.

who got a good night's sleep after training actually improved their scores when retested three days later. This suggests that the brain was continuing to learn while the person was sleeping. Students who got less than six hours of sleep after training showed no improvement or received lower scores when retested.

No one knows for sure how information is processed during sleep. Dr. Robert Stickgold of Harvard University Medical School was one of the researchers involved in the study. He theorizes that during non-REM sleep, information is passed from the hippocampus, which stores memories, to the cortex. There, proteins strengthen the connections between neurons involved in the task that was learned. During REM sleep, the brain practices the task. The result is improvement in the task when the student is retested. This research demonstrates that getting a good night's sleep after learning is important for retaining what is learned. It is another reason that adequate sleep will enhance your ability to do well in school and other activities that require knowledge and skills.

Chapter 4

Avoiding Alcohol, Tobacco, and Drugs

You've no doubt heard that you should not drink, smoke, or take drugs. This is not just to keep you from wasting money, to make you concentrate on serious activities, or to make you a better person—although these things may be true. Using alcohol, tobacco, or drugs can have serious negative effects on your brain.

Drinking and Your Brain

Drinking alcohol makes a person more likely to have a car crash or other accident that injures the head and brain. Alcohol can also harm the brain directly.

Contrary to popular belief, alcohol itself does not kill brain cells. The bad news is, it does damage them. At the ends of neurons are hairlike structures called dendrites, which neurons use to communicate with each other. A single neuron can form connections to tens of thousands of other neurons. The neurons in the brain form a network with more than one hundred trillion connections. Alcohol damages dendrites, making it difficult for neurons to communicate, which affects brain function. The damage can affect areas of the brain responsible for judgment, memory, learning, and motor function. The damage to dendrites may be reversible if a person stops drinking. However, drinking large amounts of alcohol for a long period of time

Avoiding Alcohol, Tobacco, and Drugs

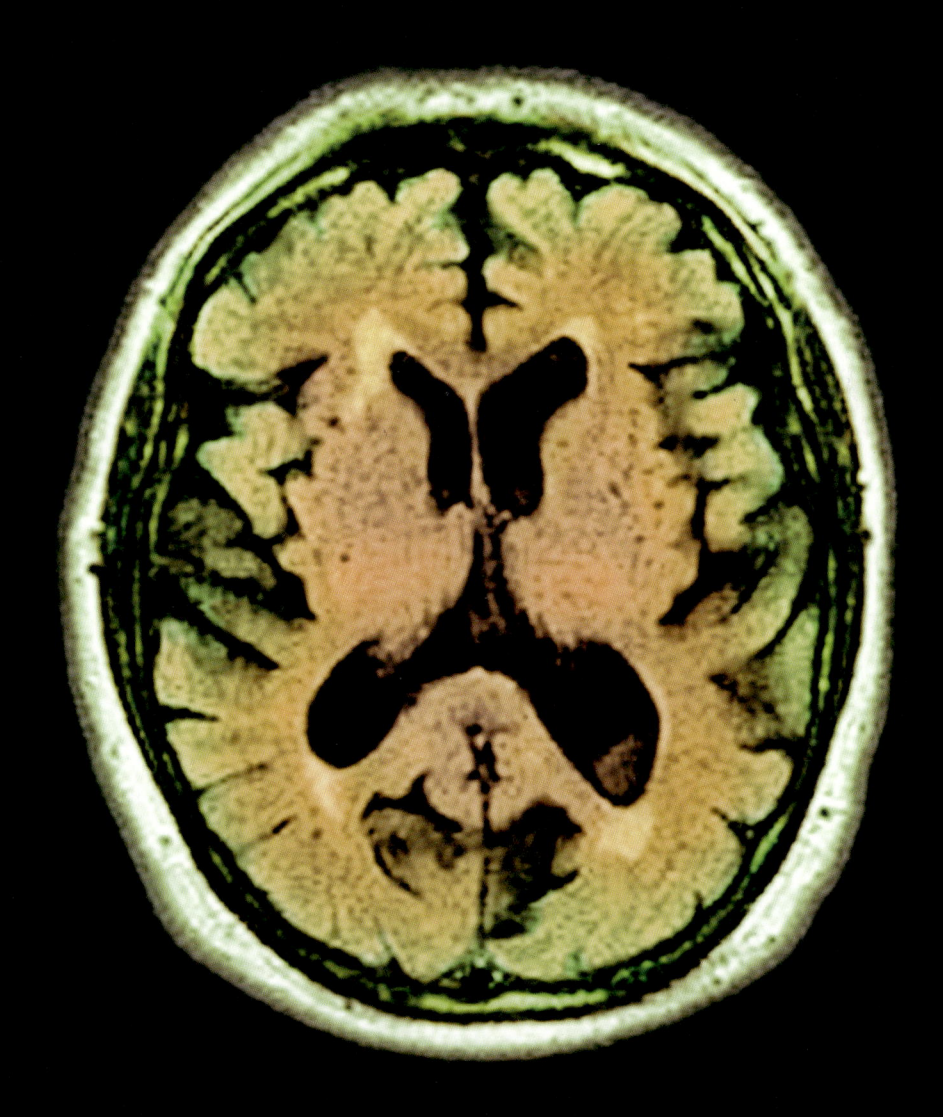

This MRI image shows how the brain tissue of an alcoholic has shrunk. Note the abnormally large spaces in the center of the brain (the ventricles) and the indentations around the outer edges.

can cause permanent changes in the brain. The best approach is not to start.

If you drink large amounts of alcohol, the problems can be even worse. Too much alcohol can cause a lack of thiamine, a B vitamin necessary for the health of neurons. Alcohol interferes with the body's ability to absorb thiamine. A deficiency of thiamine does result in the death of neurons. The resulting disorder is called Wernicke-Korsakoff syndrome. It is seen in alcoholics, and its symptoms include memory loss, confusion, lack of muscle coordination, paralysis of the eyes, and even death.

Steroids and the Brain

People do not take illegal drugs solely for the feelings they produce. Some people take steroids to build their muscles. They think bulky muscles will improve their looks or enhance their athletic performance. Steroids can be very dangerous, however. Steroids are a form of the hormone testosterone, which is responsible for the development of adult male features in boys, starting at puberty. The body's production of testosterone is controlled by the hypothalamus, located at the base of the brain. Among other things, the hypothalamus controls mood, appetite, blood pressure, and reproductive capability. Taking steroids interferes with the normal working of the hypothalamus.

In both genders, taking steroids can have negative effects. In males, taking steroids can cause shrinkage of the testes, sexual dysfunction, baldness, and rashes. Females who take steroids may find that doing so stops their periods and results in baldness, growth of facial hair, and deepening of the voice. Steroids can also affect mood. "'Roid rage" is aggressiveness and violent behavior caused by taking steroids. Because of these serious consequences, it is best to avoid taking steroids even if teammates or friends pressure you.

Effects of Drugs on the Brain

Drugs have a direct effect on the brain. The brain uses chemicals called neurotransmitters to control the functioning of neurons. Drugs are also chemicals, and they can interfere with the effects of neurotransmitters, disrupting brain function. The neurons have receptors, which are the equivalent of sockets into which certain chemicals fit, much like a key in a lock. Some drugs lock onto these receptors, causing the cells to send abnormal messages.

For example, the chemical in marijuana is called THC. The receptors to which it attaches are found in the parts of the brain relating to pleasure, memory, concentration, perception, and coordination. That is why smoking marijuana affects all these brain functions. According to the National Institute on Drug Abuse (NIDA), research studies have shown that the negative effects of marijuana on memory and learning last for days or weeks after the immediate effects have worn off. Thus, smoking marijuana regularly can result in reduced intellectual functioning on a continuing basis.

Drugs that are stimulants, such as amphetamines (speed) and cocaine, work differently. They cause the brain to release abnormally large amounts of neurotransmitters. This amplifies, or boosts, the effects of the neurotransmitters. Using such drugs initially produces a feeling of increased alertness, energy, and elevated mood. With continued use, these feelings become further exaggerated into irritability, restlessness, anxiety, and paranoia (irrational fear). Cocaine also constricts (narrows) the blood vessels in the brain, and its use can lead to seizures or stroke.

Prescription medications are usually safe if they are used to treat a specific medical condition and are taken as directed under a doctor's supervision. However, in recent years there has been an increase

Your Beautiful Brain: Keeping Your Brain Healthy

THC and the brain

Tetrahydrocannabinol (THC), the psychoactive substance found in cannabis, affects the body when marijuana is smoked or otherwise ingested. Located throughout the body, cannabinoid receptors are found in greatest quantity in the brain, particularly in areas that govern coordination, judgment, learning and memory. Some of the areas THC affects:

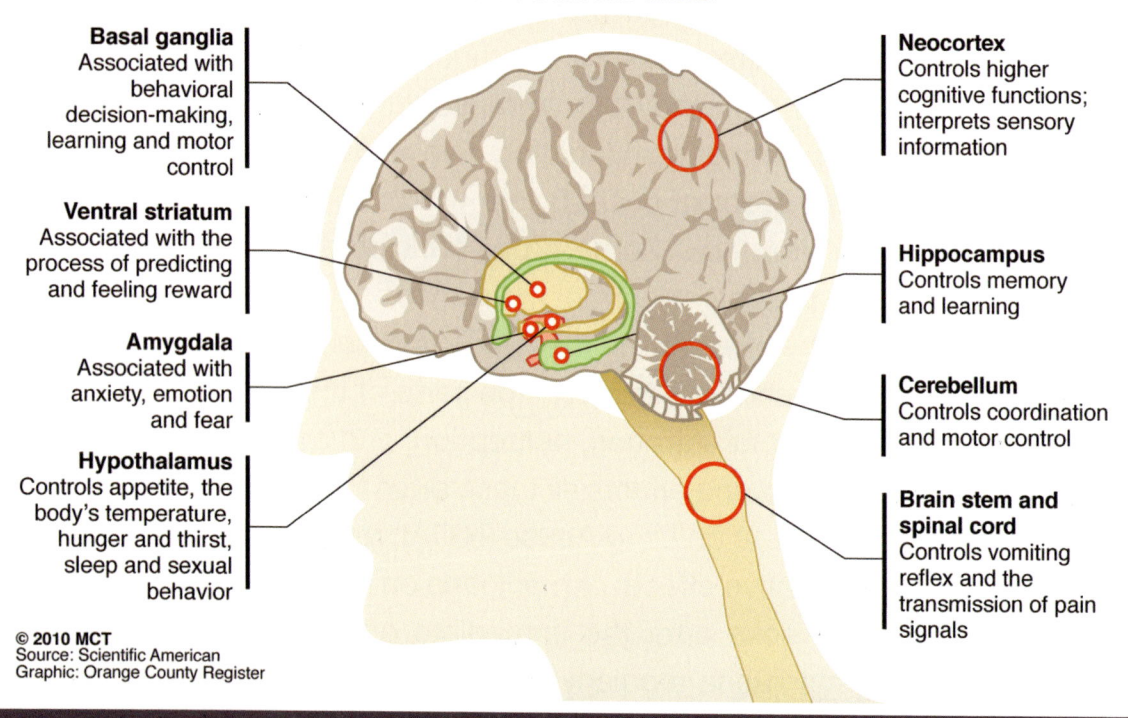

Basal ganglia
Associated with behavioral decision-making, learning and motor control

Ventral striatum
Associated with the process of predicting and feeling reward

Amygdala
Associated with anxiety, emotion and fear

Hypothalamus
Controls appetite, the body's temperature, hunger and thirst, sleep and sexual behavior

Neocortex
Controls higher cognitive functions; interprets sensory information

Hippocampus
Controls memory and learning

Cerebellum
Controls coordination and motor control

Brain stem and spinal cord
Controls vomiting reflex and the transmission of pain signals

© 2010 MCT
Source: Scientific American
Graphic: Orange County Register

This diagram of the brain shows the various areas affected by THC, the mind-altering substance found in marijuana. The effects of THC can last for days or weeks after use.

in the number of young people abusing prescription drugs. They may take prescription medicines from their parents' supply or get them from friends. Among the most commonly abused prescription drugs are narcotic painkillers, tranquilizers, and antidepressants. All of these drugs can affect brain chemistry and interfere with the normal output or uptake of neurotransmitters. It can be dangerous to take such drugs in high doses or to take them in combination. They can not only affect brain function but also slow down lung and heart function, which can be dangerous or even fatal.

Effects of Tobacco on the Brain

Smoking has a variety of negative effects on the body. It can damage the lungs, blood vessels, pancreas, stomach, and other organs. Recently, researchers have learned that smoking can also directly damage the brain. A research team at the Indian National Brain Research Center, headed by Debapriya Ghosh and Dr. Anirban Basu, discovered that a compound in tobacco causes white blood cells to attack healthy brain cells. White blood cells are part of the immune system; they normally attack and destroy bacteria and viruses that get into the bloodstream. Tobacco contains a compound known as NNK. The researchers found that NNK activates proteins in the brain that promote inflammation and signal immune system cells to attack. Brain inflammation is a feature of disorders such as multiple sclerosis (MS). In this disease, the protective covering on nerves is damaged, resulting in muscle, vision, and other problems. The research suggested that both direct and secondhand smoke might cause similar inflammatory conditions in the brain—and similar damage. So, protect your brain and the brains of others by avoiding smoking.

10 Great Questions to Ask a Health Professional

1. In what kinds of sports would I be at risk of brain injury?
2. How can I protect my brain when I engage in sports?
3. Is it safe for my brain if I engage in extreme sports?
4. How can I recognize if something is wrong with my brain?
5. What kind of diet is best to keep my brain healthy?
6. How will drinking affect my brain?
7. Is the medicine I'm taking likely to affect the development of my brain?
8. How can I improve my memory?
9. What healthy habits will help me to do well in school?
10. How can I learn more about brain function?

Chapter 5

Exercising Your Brain

The brain is an amazingly adaptable organ. When you learn, the brain forms "pathways" by creating connections between neurons. In order to do this, the brain produces a chemical called nerve growth factor (NGF) when you learn a new skill. The NGF causes dendrites to grow on neurons so that they can form connections with other neurons. This allows you to perform the new skill. Getting the brain to keep producing these chemicals and renewing connections between neurons is critical to keeping the brain healthy.

Mental exercise also benefits other structures of the brain. For example, mental activity can help keep the hippocampus healthy. Since the hippocampus is the part of the brain responsible for memory, maintaining it is crucial. The basal ganglia, clusters of nerve cells that surround the thalamus, also benefit from mental activity. The basal ganglia are responsible for physical coordination.

Learning and challenging yourself mentally helps your brain develop and also keeps your brain healthy as you age. Diseases such as Alzheimer's disease can cause severe mental decline in older people. In people without these diseases, much of the mental decline that occurs with age is the result of inactivity and a lack of brain stimulation. The more you use the brain, the better it functions. The opposite is also true. In addition, the more you learn when you are young, the better your brain will respond later in life. Research described in this chapter has shown that people who are active learners as children and teens benefit the most from mental exercise as they age.

Mental Exercises for a Brain Workout

Challenging the brain benefits brain function. The following are some mental exercises recommended by the Franklin Institute for improving brain function by learning new skills:

- Switch the hand you usually use to perform a task. For example, if you normally use your right hand to control a computer mouse or a TV remote, use your left instead. Performing the task with your other hand may be awkward at first, but learning to do so stimulates the brain.
- To improve your memory, observe objects and then try to draw them from memory. Practice this every day for a week. To exercise the part of the brain involved in long-term memory, try drawing the images you observed a week later.
- Playing games, especially strategy games like chess, is another way to challenge your brain. Doing puzzles that require thought, such as difficult crossword puzzles, can also help your brain.

Exercise your brain as well as your body. Giving your brain a difficult task to do, such as playing a complicated strategy game, keeps neurons making new connections.

Learning New Skills

When you learn a new skill, NGF spreads throughout the entire brain. It causes growth in all neurons, not just those involved in learning a particular task. Thus, learning new skills ultimately leads to the renewal of cells throughout the brain.

In order to improve brain function and keep your brain healthy, you need to learn skills that are new to you. If you know only your native language, learning a foreign language is a new skill. Learning a new pitch in baseball when you already know how to pitch is enhancing an existing skill. It will not have the same effect on the brain as

Learning a skill that requires controlling muscles in a new way, such as performing martial arts moves, can also help your brain make new connections and stay healthy.

learning a totally new skill. To benefit your brain, you must challenge it to learn activities you've never done before—learn to dance, do martial arts, or make sculpture.

Reading is another important activity for building up brain function. Dr. David Bennett of Rush University in Chicago, Illinois, says that challenging the brain at a young age is crucial to maintaining one's mental abilities later in life. Dr. Bennett told the Franklin Institute that reading habits prior to age eighteen are a key predictor of later cognitive (mental) function.

The Brain and Social Interaction

A 2010 study at the University of Michigan revealed that positive social interactions boost cognitive function. The study, described in an article on ScienceDaily.com, investigated the effects of short, positive social interactions on intellectual functioning. The study found that ten-minute positive exchanges with others boosted cognitive function in a manner equal to the effect of other types of mental exercise, such as doing crossword puzzles. Competitive exchanges with others did not provide the same benefits. More research needs to be done, but the results suggest that having a relaxed, friendly chat with a peer before a test may be a good idea.

Classic research from the late 1950s and early 1960s showed the importance of social contact for the mental development of rhesus monkeys, primates that are similar to humans. Harry and Margaret Harlow's studies of newborn rhesus monkeys showed that the monkeys did not develop normal cognitive skills if they were isolated from their mothers and other live monkeys.

Today we are still learning about how social interaction affects the brain and why it is so important. According to a report from the Society

Positive social interaction can improve brain function. Studying with friends and supporting each other's efforts in a friendly way can help you perform better.

for Neuroscience, both monkeys and humans appear to have a "mirror neuron" system. This means the neurons "fire" not only when an individual performs an action or movement but also when an individual observes someone else perform the same action. For example, whether you see someone else smile or are smiling yourself, the same parts of the brain become activated. Future research will likely explore the role of the mirror neuron system in developing empathy for other people, learning by observation, and gaining language skills.

There is no question that positive face-to-face social interaction is beneficial to the brain. What about social networking—electronic contact through Web sites such as Facebook and Twitter? The jury is still out. Some experts feel that positive contact through social networking sites helps young people develop empathy as they support their friends. Other experts feel that communicating at a distance depersonalizes communication, making young people less sensitive to others and more self-centered. Dr. Marco Iacoboni, a neuroscientist at the University of California, Los Angeles, told the *New York Times* that mirror neurons work best when people are face-to-face. Remember, electronic communication is no substitute for real-life human interaction.

You now know a variety of mental, physical, and social habits that will keep your brain healthy. The rest is up to you.

GLOSSARY

adrenal glands A pair of organs that sit on the kidneys and produce hormones, including adrenaline.

basal ganglia Clusters of nerve cells surrounding the thalamus that control movement.

blood-brain barrier A layer of tightly packed cells in the blood vessels of the brain that keep substances in the blood from passing into the brain.

body mass index (BMI) A measure consisting of a comparison of height to weight that is used to determine whether a person is obese or underweight.

Broca's area An area of the brain that controls movement necessary for speech.

circulatory system The heart and network of blood vessels in the body.

cognitive Relating to mental activities such as perception, reasoning, and judgment.

concussion A traumatic brain injury caused by jarring of the brain. Physical symptoms include headaches and dizziness. Changes in brain function, such as memory loss and attention problems, can also occur.

cortisol A hormone released in response to stress that prepares the body to defend itself.

dementia Mental decay and loss of brain function.

dendrite A hairlike fiber on the end of a neuron that receives communication from other neurons.

gene A piece of DNA that carries hereditary information.

genetic Relating to genes and inherited characteristics.

Your Beautiful Brain: Keeping Your Brain Healthy

hormone A chemical produced in the body that controls organ function.

inert Inactive.

inflammation The reaction of a body part to injury or infection that includes increased blood flow with an influx of white blood cells.

magnetic resonance imaging (MRI) A technology that uses large magnets to create images of the body.

molecule The smallest unit of a compound or element.

motor skill Controlled movement of muscles in a learned sequence in order to perform a task.

mutation A change in a code for a gene.

neuron A nerve cell that is the basic unit of the brain and nervous system and works to process and transmit information.

neurotransmitter A chemical that carries messages from one neuron to another.

obesity The condition of being extremely overweight.

paralysis A loss of movement or sensation in a part of the body.

peripheral nerve A nerve that lies outside of the brain and spinal cord.

protein A compound made up of a long chain of amino acids. Proteins are essential for life and are a part of all living cells.

puberty The stage of adolescence marked by physical growth and sexual maturation.

retention The ability to keep things in the mind and remember them for a long time.

secrete To produce and release a product from a cell or gland.

seizure A sudden attack of abnormal brain activity, resulting in violent muscle contractions, sensory symptoms, or loss of consciousness.

GLOSSARY

sleep deprivation A lack of the necessary amount of sleep over a period of time. It can cause physical or mental symptoms and affect the performance of everyday tasks.

stimulant A substance that causes increased bodily activity, especially in the nervous and circulatory systems. A stimulant can make a person feel excited or invigorated.

stroke The bursting or blockage of a blood vessel in the brain.

synapse The junction or gap where nerve impulses pass from one neuron to another or from a neuron to a muscle cell or gland cell.

traumatic brain injury (TBI) Damage to the brain that results from an injury, usually a violent blow or jolt to the head.

Wernicke's area The area of the brain responsible for comprehending speech.

FOR MORE INFORMATION

Alzheimer Society of Canada
20 Eglinton Avenue West, Suite 1600
Toronto, ON M4R 1K8
Canada
(416) 488-8772
Web site: http://www.alzheimer.ca

The Alzheimer Society of Canada provides support, information, and education to people with Alzheimer's disease and related dementias, promotes research, and leads the search for a cure. The organization provides information on how to improve and protect your brain health.

Brain Injury Association of America (BIAA)
1608 Spring Hill Road, Suite 110
Vienna, VA 22182
(703) 761-0750
Web site: http://www.biausa.org

BIAA is the country's oldest and largest brain injury advocacy organization. Through advocacy, education, and research, BIAA helps people living with brain injury, their families, and the professionals who serve them.

The Dana Foundation
505 Fifth Avenue, 6th Floor
New York, NY 10017
(212) 223-4040
Web site: http://www.dana.org

FOR MORE INFORMATION

The Dana Foundation supports brain research through grants and educates the public about the successes and potential of brain research. The organization produces free publications; coordinates the International Brain Awareness Week campaign; supports the Dana Alliances, a network of neuroscientists; and maintains an informative Web site.

National Institute on Drug Abuse (NIDA)
6001 Executive Boulevard, Room 5213
Bethesda, MD 20892-9561
(301) 443-1124
Web site: http://drugabuse.gov
NIDA conducts research and distributes the latest scientific information about drug abuse and addiction to drug abuse researchers, health professionals, teachers, advocacy groups, teenagers, and the general public.

National Institute of Mental Health (NIMH)
Science Writing, Press, and Dissemination Branch
6001 Executive Boulevard, Room 8184, MSC 9663
Bethesda, MD 20892-9663
(866) 615-6464
Web site: http://www.nimh.nih.gov
NIMH seeks to reduce suffering from mental illness and behavioral disorders through research on the mind, brain, and behavior. NIMH provides a wide range of information based on that research.

National Institute of Neurological Disorders and Stroke (NINDS)
P.O. Box 5801
Bethesda, MD 20824
(800) 352-9424
Web site: http://www.ninds.nih.gov
NINDS conducts and supports research on the causes, prevention, diagnosis, and treatment of neurological disorders and stroke.

National Sleep Foundation (NSF)
1010 N. Glebe Road, Suite 310
Arlington, VA 22201
(703) 243-1697
Web site: http://www.sleepfoundation.org
NSF is dedicated to improving the quality of life for Americans who suffer from sleep problems and disorders. The organization helps the public better understand the importance of sleep and the benefits of good sleep habits. It also helps people recognize the signs of sleep problems so that they can be properly diagnosed and treated.

President's Council on Fitness, Sports and Nutrition (PCFSN)
1101 Wootton Parkway, Suite 560
Rockville, MD 20852
(240) 276-9567
Web sites: http://www.fitness.gov; http://www.presidentschallenge.org
PCFSN's mission is to engage, educate, and empower all Americans to adopt a healthy lifestyle that includes regular physical activity and good nutrition.

FOR MORE INFORMATION

Society for Neuroscience (SfN)
1121 14th Street NW, Suite 1010
Washington, DC 20005
(202) 962-4000
Web site: http://www.sfn.org

The Society for Neuroscience (SfN) is an organization of scientists and physicians dedicated to analyzing the nervous system and its role in everything we do. SfN works to advance research on the nervous system, leading to a better understanding of how the brain works and new ways to treat nervous system disorders. It also works to inform the public about the progress and benefits of neuroscience research and to promote education in this field.

Web Sites

Due to the changing nature of Internet links, Rosen Publishing has developed an online list of Web sites related to the subject of this book. This site is updated regularly. Please use this link to access the list:

http://www.rosenlinks.com/hab/brain

FOR FURTHER READING

Bangalore, Lakshmi. *Brain Development* (Gray Matter). New York, NY: Chelsea House, 2007.

Capaccio, George. *Nervous System* (Amazing Human Body). New York, NY: Marshall Cavendish Benchmark, 2010.

Carter, Rita. *The Human Brain Book*. New York, NY: DK Publishing, 2009.

Hoffelder, Ann McIntosh, and Robert L. Hoffelder. *How the Brain Grows* (Brain Works). New York, NY: Chelsea House, 2007.

Markle, Sandra. *Wounded Brains: True Survival Stories* (Powerful Medicine). Minneapolis, MN: Lerner Publications, 2010.

Parker, Steve. *Brain: Injury, Illness, and Health* (Body Focus). Chicago, IL: Heinemann, 2009.

Rau, Dana Meachen. *Freaking Out! The Science of the Teenage Brain* (Everyday Science). Mankato, MN: Compass Point Books, 2012.

Rodriguez, Ana Maria. *A Day in the Life of the Brain* (Brain Works). New York, NY: Chelsea House, 2007.

Shea, Therese. *Dementia* (Understanding Brain Diseases and Disorders). New York, NY: Rosen Publishing, 2012.

Simon, Seymour. *The Brain: Our Nervous System*. New York, NY: Collins, 2006.

Snedden, Robert. *Understanding the Brain and the Nervous System* (Understanding the Human Body). New York, NY: Rosen Central, 2010.

Stewart, Melissa. *You've Got Nerve! The Secrets of the Brain and Nerves* (The Gross and Goofy Body). New York, NY: Marshall Cavendish Benchmark, 2011.

FOR FURTHER READING

Stewart, Sheila, and Camden Flath. *What's Wrong with My Brain? Kids with Brain Injury*. Broomall, PA: Mason Crest Publishers, 2011.

Stimola, Aubrey. *Brain Injuries* (Understanding Brain Diseases and Disorders). New York, NY: Rosen Publishing, 2012.

Vera-Portocarrero, Louis. *Brain Facts* (Gray Matter). New York, NY: Chelsea House, 2007.

Woodward, John, Serge Seidlitz, and Andy Smith. *How to Be a Genius*. New York, NY: DK Publishing, 2009.

BIBLIOGRAPHY

Associated Press. "Can Folic Acid Slow Brain Drain?" CBSNews.com, February 11, 2009. Retrieved September 3, 2011 (http://www.cbsnews.com/stories/2005/06/20/health/main703140.shtml).

Aubrey, Allison. "A Better Breakfast Can Boost a Child's Brainpower." NPR.org, August 31, 2006. Retrieved September 5, 2011 (http://www.npr.org/templates/story/story.php?storyId=5738848).

Blakeslee, Sandra. "Cells that Read Minds." *New York Times*, January 10, 2006. Retrieved September 5, 2011 (http://www.nytimes.com/2006/01/10/science/10mirr.html?pagewanted=all).

Centers for Disease Control and Prevention. "Obesity and Overweight for Professionals: Data and Statistics: U.S. Obesity Trends." July 21, 2011. Retrieved September 3, 2011 (http://www.cdc.gov/obesity/data/trends.html).

Fernandez, Alvaro, and Elkhonon Goldberg. *The SharpBrains Guide to Brain Fitness*. San Francisco, CA: SharpBrains, Inc., 2009.

The Franklin Institute Online. "The Human Brain—Exercise." 2004. Retrieved September 1, 2011 (http://www.fi.edu/learn/brain/exercise.html#mentalexercise).

The Franklin Institute Online. "The Human Brain—Stress." 2004. Retrieved September 3, 2011 (http://www.fi.edu/learn/brain/stress.html).

Howard, Pierce J. *The Owner's Manual for the Brain: Everyday Applications from Mind-Brain Research*. 3rd ed. Austin, TX: Bard Press, 2006.

HowStuffWorks.com. "The Brain and Nervous System." May 3, 2006. Retrieved September 1, 2011 (http://health.howstuffworks.com/human-body/systems/nervous-system/brain-nervous-system-ga.htm).

BIBLIOGRAPHY

National Institute on Alcohol Abuse and Alcoholism. "Alcohol's Damaging Effects on the Brain." *Alcohol Alert*, No. 63, October 2004. Retrieved September 5, 2011 (http://pubs.niaaa.nih.gov/publications/aa63/aa63.htm).

National Institute on Drug Abuse. "Drugs, Brains, and Behavior—The Science of Addiction." August 2010. Retrieved September 26, 2011 (http://drugabuse.gov/scienceofaddiction/brain.html).

National Institute on Drug Abuse. "NIDA for Teens: Mind Over Matter—Anabolic Steroids." NIH Publication No. 03-3860, 2003. Retrieved September 4, 2011 (http://teens.drugabuse.gov/mom/mom_ster1.php).

National Institute on Drug Abuse. "NIDA InfoFacts: Marijuana." November 2010. Retrieved September 26, 2011 (http://www.nida.nih.gov/infofacts/marijuana.html).

National Institute of Mental Health. "Teenage Brain: A Work in Progress." NIH Publication No. 01-4929, June 26, 2008. Retrieved September 1, 2011 (http://wwwapps.nimh.nih.gov/health/publications/teenage-brain-a-work-in-progress.shtml).

National Institute of Neurological Disorders and Stroke. "Brain Basics: Understanding Sleep." NIH Publication No.06-3440-c, May 21, 2007. Retrieved September 3, 2011 (http://www.ninds.nih.gov/disorders/brain_basics/understanding_sleep.htm).

National Institute of Neurological Disorders and Stroke. "NINDS Traumatic Brain Injury Information Page." April 15, 2011. Retrieved September 4, 2011 (http://www.ninds.nih.gov/disorders/tbi/tbi.htm).

National Library of Medicine. "Alzheimer's Disease." MedlinePlus, July 13, 2011. Retrieved September 2, 2011 (http://www.nlm.nih.gov/medlineplus/alzheimersdisease.html).

National Library of Medicine. "Brain Diseases." MedlinePlus, May 15, 2011. Retrieved September 2, 2011 (http://www.nlm.nih.gov/medlineplus/braindiseases.html).

National Library of Medicine. "Stroke." MedlinePlus, 2011. Retrieved September 2, 2011 (http://www.nlm.nih.gov/medlineplus/stroke.html).

National Library of Medicine. "Traumatic Brain Injury." MedlinePlus, August 24, 2011. Retrieved September 5, 2011 (http://www.nlm.nih.gov/medlineplus/traumaticbraininjury.html).

National Sleep Foundation. "The ABCs of ZZZZs—When You Can't Sleep." Retrieved September 4, 2011 (http://www.sleepfoundation.org/article/how-sleep-works/abcs-zzzzs-when-you-cant-sleep).

Society for Neuroscience. "Mirror Neurons." November 2008. Retrieved September 5, 2011 (http://www.sfn.org/index.aspx?pagename=brainbriefings_mirrorneurons).

Spinks, Sarah. "From Zzzz's To A's—Adolescents and Sleep: Inside the Teenage Brain." PBS.org. Retrieved September 4, 2011 (http://www.pbs.org/wgbh/pages/frontline/shows/teenbrain/from/sleep.html).

Spinks, Sarah. "Work in Progress—One Reason Teens Respond Differently to the World: Inside the Teenage Brain." PBS.org. Retrieved September 4, 2011 (http://www.pbs.org/wgbh/pages/frontline/shows/teenbrain/work/onereason.html).

Swaminathan, Nikhil. "Why Does the Brain Need So Much Power?" *Scientific American*, April 29, 2008. Retrieved September 3, 2011 (http://www.scientificamerican.com/article.cfm?id=why-does-the-brain-need-s).

University of California, Los Angeles. "Scientists Learn How Food Affects the Brain: Omega 3 Especially Important." ScienceDaily

BIBLIOGRAPHY

.com, July 9, 2008. Retrieved September 3, 2011 (http://www.sciencedaily.com/releases/2008/07/080709161922.htm).

University of Michigan. "Friends with Cognitive Benefits: Mental Function Improves After Certain Types of Socializing." ScienceDaily.com, October 28, 2010. Retrieved September 12, 2011 (http://www.sciencedaily.com/releases/2010/10/101028113817.htm).

Wiley-Blackwell. "Smoking Linked to Brain Damage." ScienceDaily.com, June 23, 2009. Retrieved September 4, 2011 (http://www.sciencedaily.com/releases/2009/06/090623090400.htm).

INDEX

A

alcohol, 5, 18, 31, 32, 36–38, 42
Alzheimer's disease, 14, 23, 43
amygdala, 9, 15–16
antioxidants, 25
arteriosclerosis, 14, 24
axon, 13

B

brain
   diseases, 14
   function, 4, 6, 8–9
   myths and facts about, 17
   structure, 8–13
   teenage, 13, 15–16
brain stem, 8–9
Broca's area, 11

C

cell body, 11
cerebellum, 8, 9
cerebral cortex, 10
cerebrum, 8
   structure, 9–11
concussion, 28
corpus callosum, 9
cortisol, 29

D

dendrites, 13
drugs, 5, 31, 36, 38, 39–40

E

emotions, 4, 6, 9, 15–16, 29
exercise
   mental, 5, 43–44
   physical, 18, 25–27, 31, 32, 42

F

frontal lobes, 10, 11, 15

G

glial cell, 6

H

helmets, 28
high blood pressure, 14, 24, 25
hippocampus, 9, 29, 35, 43
hormones, 9, 33, 38
Huntington's disease, 14
hypothalamus, 9, 38

L

learning, 11, 20, 21, 25, 30, 33–35, 39, 43, 44, 45–46
limbic system, 8, 9

M

magnetic resonance imaging (MRI), 13, 15
medulla, 8
memory, 4, 6, 9, 19, 20, 21, 23, 25, 30, 33, 38, 39, 42, 43, 44
midbrain, 8
multiple sclerosis, 41

N

neurons, 6, 18, 19, 23, 25, 33, 36, 38, 39, 43, 47, 48
   structure, 11–13
neurotransmitters, 13, 18, 23, 39, 40
nutrition, 5, 18–19, 42
   breakfast, 21–23
   fats and sugars, 19–21
   for protecting the brain, 23, 25

INDEX

O
obesity, 21, 24–25
occipital lobes, 10, 11

P
parietal lobes, 10, 11, 15
peripheral nerves, 6
pituitary gland, 9
pons, 8

S
sleep, 5, 9, 29, 31–35
social interaction, 46–48

spinal cord, 6
steroids, 38
stress, 29–31, 33
stroke, 14, 24, 25, 39
synapse, 13, 19, 25, 27

T
temporal lobes, 10, 11, 15
thalamus, 9, 43
tobacco, 5, 14, 32, 36, 41
traumatic brain injuries, 27–28

W
Wernicke's area, 11

About the Author

Jeri Freedman has a B.A. from Harvard University. For fifteen years she worked for companies in the medical field. She is the author of more than thirty young adult nonfiction books, many published by Rosen Publishing. Among her previous titles are *How Do We Know About Genetics and Heredity*, *The Mental and Physical Effects of Obesity*, and *Steroids: High-Risk Performance Drugs*.

Photo Credits

Cover © istockphoto.com/digitalskillet; p. 4 Thinkstock/Comstock/Thinkstock; p. 7 Sebastian Kaulitzki/Shutterstock; p. 8 udaix/Shutterstock; p. 10 MedusArt/Shutterstock; p. 12 ducu59us/Shutterstock; p. 19 BananaStock/Thinkstock; p. 22 Hemera/Thinkstock; p. 24 Diamond_Images/Shutterstock; pp. 26, 30 samotrebizan/Shutterstock; p. 34 istockphoto/Thinkstock; p. 37 Du Cane Medical Imaging Ltd./Photo Researchers, Inc.; p. 40 Staff/MCT/Newscom; p. 44 AVAVA/Shutterstock; p. 45 Hemera Technologies/Ablestock.com/Thinkstock; p. 47 Monkey Business Images/Shutterstock; cover (background graphic), interior graphic (frame) © istockphoto.com/liquidplanet; interior graphic (ECG waves) © istockphoto.com/linearcurves.

Designer: Nicole Russo; Editor: Andrea Sclarow Paskoff
Photo Researcher: Marty Levick